INTRODUCTION

This booklet by the Astronomy Correspondent of *The Times*, Michael J. Hendrie, provides a convenient guide to seeing the stars, Moon and planets with the naked eye. The twelve monthly charts on pp.4–26 show those objects above the horizon late in the evening. The charts have been drawn for the latitude of London (51°30' north) but may be used for any part of the British Isles. Opposite each chart are notes on the visibility of the planets and phases of the Moon. Times of sunset, twilight and sunrise are tabulated on page 29 along with notes on the principal meteor showers. Further notes on the more interesting events in 2002 appear on pp.31–32. A detailed explanation of the astronomical terms used and how the various phenomena arise can be found in the fully illustrated *The Times Night Sky Companion*.

The Changing Aspect of the Night Sky

From our position on the surface of the Earth, the stars appear to lie on the inside of a spherical surface, called the celestial sphere. Because the stars are so far away their directions remain essentially unchanged when seen from different parts of the Earth's orbit. The diagram on the inside cover *(see* left) shows one such direction, indicated by the arrow pointing to the First Point of Aries. The stars that lie behind the Sun as seen from the Earth at the beginning of April each year will, by October, be in the opposite part of the sky to the Sun and be due south at midnight. The arrow points to Pisces (where the First point of Aries now lies) and looking at the October chart Pisces is indeed in the southern sky but it does not appear at all on the April chart, being behind the Sun and in the daytime sky. More generally, a line from the Sun through the position of the Earth points towards the stars seen near the lower centre of the chart for that month. Remembering this, one can relate the positions of the other planets in their orbits to where they will be in the sky, though they will not always be on the monthly chart, being too near to the Sun in direction.

Time of Observation and Location

Greenwich Mean Time (GMT but also known as Universal Time) is used throughout this booklet. When in force, British Summer Time (BST) is 1 hour ahead of GMT, e.g. 23h BST is 22h GMT. Strictly speaking, the charts are only correct in the stars they show above the horizon for an observer near London (Greenwich). As one moves north fewer stars appear above the southern horizon. Movement east or west along the same latitude does not alter what stars can be seen, but only when they can be seen.

Using the Charts

The charts show the brighter stars above the horizo͏ͬ ͫ ͭ ͦ ͫ ͩ ͨ ͫ ͣ ͭ 23h (11pm) at the beginning, 22h (10pm) in the middle and 21h (9p͏ stars rise four minutes earlier each night or two ho͏ in their same positions at the same time after a ye͏ the heavens at 23h on 1 April is the same as on 1

remembering this rule, the chart applicable to any hour throughout the year may be found. This rule does not apply to the Moon and planets. The charts show the whole sky visible at one time with the zenith, the point directly overhead, at the centre of the chart. Note that the Pole Star (Polaris) occupies the same position on every chart being close to one of the two points around which the whole star sphere appears to revolve. It is easily found in relation to Ursa Major at all times of the year and is useful in defining due north. Ursa Major's seven brightest stars form the Plough. The end two stars (the Pointers) are always in line with Polaris. If the observer faces south with the Pole Star to his back and the appropriate chart held up as one would read the booklet, the constellations depicted above the southern horizon should be to the front, with the eastern aspect to the left and western horizon to the right.

Explanatory Notes on Terms Used

The Moon – the phase and position are given for about 22h on every other day when it is above the horizon at that time. The average time between like phases (e.g. full to full) is 29.5 days, 2 days longer than it takes to return amongst the same stars. It moves eastwards by its own diameter every hour.

The Planets – are shown in the position they occupy about the middle of the month unless otherwise indicated, and for Venus and Mars an arrow shows by its length the movement during the month. Planets crossing the meridian (i.e. due south) before midnight are said to be evening stars while those crossing the meridian after midnight are morning stars. A planet is in opposition to the Sun when it is in the opposite part of the sky to the Sun and therefore due south at midnight. (Mercury and Venus can never be at opposition.) It is then at its closest and brightest for that year. For a few weeks on either side of opposition, motion among the stars, instead of being from west to east as usual, is from east to west and is called retrograde. At the turning points, where motion is reversed, the planet is said to be stationary. A planet coming in line with the Earth and the Sun is said to be in superior conjunction with the Sun if it lies beyond the Sun but at inferior conjunction if it lies between the Earth and the Sun. Only Mercury and Venus can be at inferior conjunction. Planets can also be in conjunction with others when close in the sky. Mercury and Venus are said to be at greatest elongation when at their greatest apparent distance from the Sun, either east (evening) or west (morning). They can never be high in the sky late at night. Mercury is not observable in a dark sky from the British Isles and may require binoculars. It is always too near the sun to be included on the monthly charts. Uranus is visible at times to the naked eye but will probably require binoculars for identification. Neptune always requires optical aid. Pluto requires a moderate-sized telescope and is not mentioned in the monthly notes. Opposition in 2002 is on 7 June, the 14th magnitude planet being in Ophiuchus.

ECLIPSES IN 2002

26 MAY

This penumbral eclipse of the Moon will be visible from the Americas (except the NE), Pacific Ocean, Asia, Australia and most of Antarctica. In a penumbral eclipse the Moon passes only through the Earth's outer shadow and is only slightly darkened. Many penumbral eclipses pass unnoticed.

10–11 JUNE

The track of this annular eclipse of the Sun stretches across the N Pacific Ocean from Indonesia in the west to just touch Mexico in the east. A partial eclipse of the Sun will be visible over a wide area covering the Philippines, Japan, mainland E Asia and N America (except the extreme E) to the north and Indonesia and NE Australia to the south.

24 JUNE

This penumbral eclipse of the Moon will be visible over a wide area from S and W Asia, the Indian Ocean, Africa, Europe and much of the N and S Atlantic Ocean. From the British Isles the Moon will be low in the SE at mid eclipse (21h 27m) and even when the Moon leaves the penumbra (22h 35m) there will still be twilight, so the eclipse may not be very noticeable.

19–20 NOVEMBER

The third penumbral lunar eclipse of the year will be visible from Africa, Europe, the Arctic, much of the Americas and the E Pacific Ocean. The Moon enters the penumbra on the 19th at 23h 32m, mid eclipse is at 1h 46m and the eclipse ends at 4h 01m. As the Moon will be high in the sky from the British Isles, the slight darkening of the Moon may be noticeable.

4 DECEMBER

The path of this total solar eclipse begins in the E Atlantic Ocean crosses southern Africa and the S Indian Ocean to end in southern and central Australia. A partial solar eclipse will be visible over central and E Africa, the S Indian Ocean, a part of Indonesia, Australia and part of Antarctica. In Africa totality will be in the morning but in Australia it will occur close to sunset.

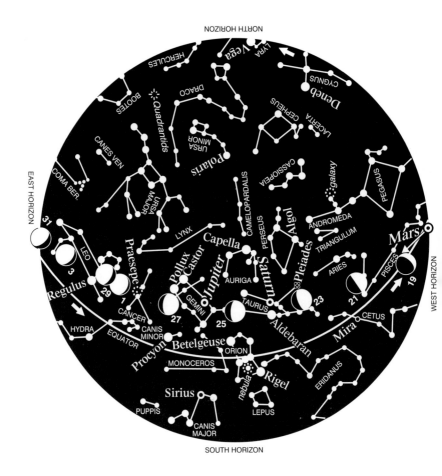

JANUARY 1, 23h (11pm)

The aspect of the sky (apart from the Moon and Planets)
will be approximately the same in other months at the
following times:

October 1, 05h: November 1, 03h: December 1, 01h: February 1, 21h: March 1, 19h.

The time in these notes is that of the Greenwich meridian.

4

JANUARY

The Planets

MERCURY is 0.0 magnitude and at greatest eastern elongation (19 degrees) on the 11th. Visible low in the SW after sunset during mid-month. It reaches inferior conjunction on the 27th, becoming a morning star though not observable.

VENUS is not visible this month moving from the morning into the evening sky at superior conjunction on the 14th.

MARS is 0.9 magnitude and passes from Aquarius into in Pisces setting about 22h all the month. It is shown on the chart from 15th–31st. Moon to the south on the 18th.

JUPITER is –2.6 magnitude and in Gemini. At opposition on the 1st when in crosses the meridian at midnight. Moon nearby on the 26th.

SATURN is –0.2 magnitude and in Taurus, setting by 03h 30m on the 31st. Moon nearby on 24th.

URANUS is in Capricornus setting soon after the Sun by the 31st.

NEPTUNE is in Capricornus and in conjunction with the Sun on the 28th and will not be observable.

The Moon

Last quarter 6d 04h
New Moon 13d 13h
First quarter 21d 18h
Full Moon 28d 23h

The Earth: at perihelion 2d 14h (147 million km)

FEBRUARY 1, 23h (11pm)

The aspect of the sky (apart from the Moon and Planets) will be approximately the same in other months at the following times:

November 1, 05h: December 1, 03h: January 1, 01h: March 1, 21h: April 1, 19h.

The time in these notes is that of the Greenwich meridian.

FEBRUARY

The Planets

MERCURY reaches greatest western elongation (27 degrees) on the 21st, but being south of the Sun, will not be visible this month.

VENUS sets an hour after the Sun by the end of February when it may just be visible very low in the SW.

MARS is 1.2 magnitude, moving from Pisces into Aries late in the month. It sets after 22h throughout the month. Positions shown for 14th–28th. Crescent Moon to the south on the 16th.

JUPITER is –2.5 magnitude and in Gemini setting by 04h on the 28th. Moon nearby on the 22nd.

SATURN is –0.1 magnitude, in Taurus and stationary on the 8th. It sets before 02h by the 28th. Moon nearby on the 20th.

URANUS is in conjunction with the Sun on the 13th and will not be observable.

NEPTUNE is in Capricornus and rises about 05h 40m by end February.

The Moon

Last quarter 4d 14h
New Moon 12d 08h
First quarter 20d 12h
Full Moon 27d 09h

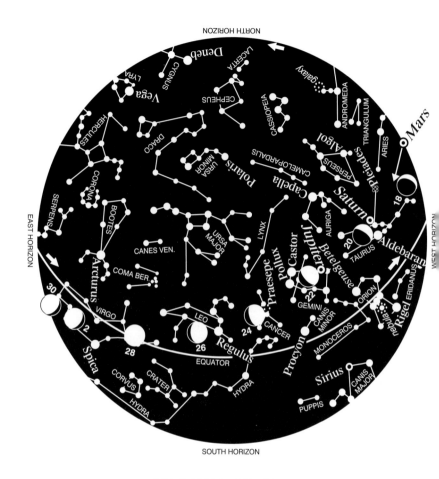

MARCH 1, 23h (11pm)

The aspect of the sky (apart from the Moon and Planets) will be approximately the same in other months at the following times:

December 1, 05h: January 1, 03h: February 1, 01h: April 1, 21h.

The time in these notes is that of the Greenwich meridian.

MARCH

The Planets

MERCURY is a morning star throughout March but is too near the Sun for observation.

VENUS is in the western evening sky setting about 2h after the Sun by the 31st.

MARS is 1.3 magnitude and in Aries, setting after 22h through March. Shown on the chart 15th–31st. Moon nearby on the 17th.

JUPITER is in Gemini at –2.3 magnitude and stationary on the 1st. It sets about 02h on the 31st. Moon nearby on the 21st–22nd.

SATURN is 0.1 magnitude and in Taurus, setting about 0h by the end of the month. Moon nearby on the 19th–20th.

URANUS moves from Capricornus into Aquarius in late March but is still in morning twilight. Moon to the north on the 11th–12th.

NEPTUNE is in Capricornus rising before 04h by the 31st. Moon to the north on the 10th.

The Moon

Last quarter 6d 01h
New Moon 14d 02h
First quarter 22d 02h
Full Moon 28d 18h

The Earth: Spring Equinox 20d 19h

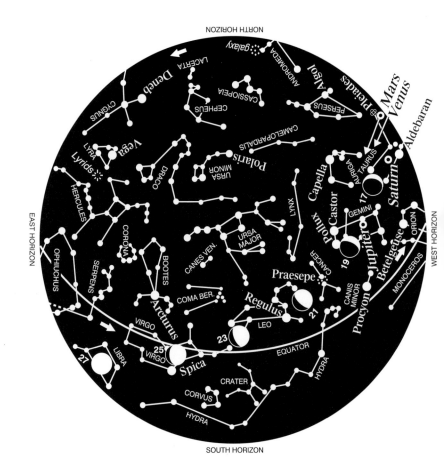

APRIL 1, 23h (11pm)

The aspect of the sky (apart from the Moon and Planets) will be approximately the same in other months at the following times:

December 1, 07h: January 1, 05h: February 1, 03h: March 1, 01h: May 1, 21h.

The time in these notes is that of the Greenwich meridian.

APRIL

The Planets

MERCURY is at superior conjunction on the 7th. The best evening apparition of the year, it can be seen in the NW from the 21st. It fades slowly to 0.0 magnitude by the 30th when it sets 2h after the Sun.

VENUS is a brilliant −3.9 magnitude in the NW evening sky during April, setting after 21h 30m by the 30th. Moon to the south on the 14th.

MARS is 1.5 magnitude, passing from Aries into Taurus in early April. It sets after 22h throughout the month. Moon to the south on the 15th.

JUPITER is −2.1 magnitude in Gemini and sets soon after 0h by the 30th. Moon close by on the 18th.

SATURN is in Taurus and sets about 22h by the 30th. Moon close by on the 16th.

URANUS is in Aquarius rising about 02h 30m by the 30th. Moon to the north on the 8th.

NEPTUNE is in Capricornus rising at 02h by the 30th. Moon nearby on the 6th.

The Moon

Last quarter 4d 15h
New Moon 12d 19h
First quarter 20d 13h
Full Moon 27d 03h

Planetary conjunctions: *see* page 32

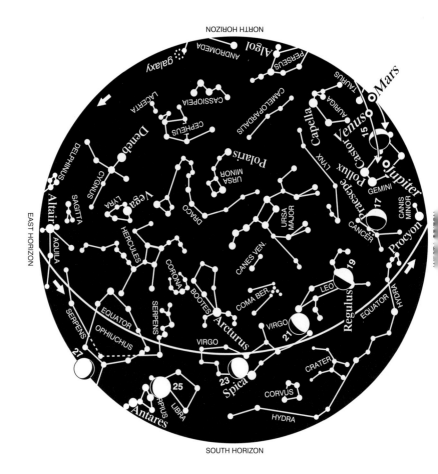

MAY 1, 23h (11pm)

The aspect of the sky (apart from the Moon and Planets) will be approximately the same in other months at the following times:

January 1, 07h: February 1, 05h: March 1, 03h: April 1, 01h: June 1, 21h.

The time in these notes is that of the Greenwich meridian.

MAY

The Planets

MERCURY begins the month as a 0.0 magnitude object in the NW evening sky reaching greatest eastern elongation (21 degrees) on the 4th. It closes rapidly with the Sun by mid-May to be at inferior conjunction by the 27th. It then becomes a morning object but is too near the Sun for observation.

VENUS is –3.9 magnitude and in the NW evening sky, setting shortly before 23h by the 31st.
Saturn to the south on the 7th, Mars nearby on the10th, Moon nearby on the 14th.

MARS is 1.7 magnitude, passing from Taurus into Gemini, setting by 22h by the 31st. Positions for 15th to 31st on the chart. North of Saturn on the 4th. Moon nearby on the 14th.

JUPITER is –1.9 magnitude and in Gemini, setting before 23h by the 31st. Moon nearby on the 16th.

SATURN is 0.1 magnitude and in Taurus, setting soon after the Sun by the 31st. Moon nearby on the 14th.

URANUS is in Aquarius rising before 01h by the 31st. Moon to the south on the 5th.

NEPTUNE is in Capricornus rising by 0h in late May. Stationary on the 13th. Moon to the south on the 4th and 31st.

The Moon

Last quarter 4d 07h
New Moon 12d 11h
First quarter 19d 20h
Full Moon 26d 12h

Eclipse on the 26th: *see* page 3
Planetary conjunctions: *see* page 32

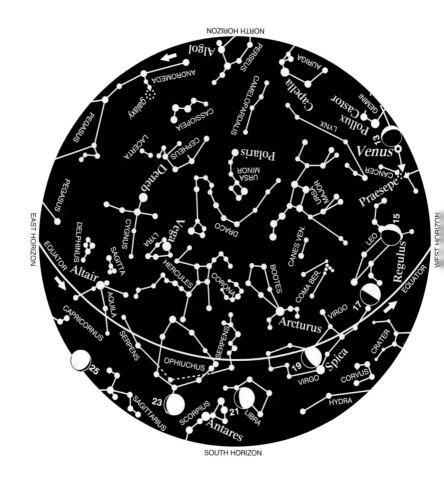

JUNE 1, 23h (11pm)

The aspect of the sky (apart from the Moon and Planets) will be approximately the same in other months at the following times:

**February 1, 07h: March 1, 05h: April 1, 03h:
May 1, 01h: July 1, 21h.**

The time in these notes is that of the Greenwich meridian.

JUNE

The Planets

MERCURY reaches greatest western elongation (23 degrees) on the 21st and might just be glimpsed in strong morning twilight in the NE late in the month at 0.0 magnitude.

VENUS is –4.0 magnitude, a brilliant object in the NW sky throughout the month, setting just before 23h. Jupiter to the south on the 3rd. Moon close by on the 13th.

MARS is near its faintest at 1.7 magnitude and is in Gemini as it runs into evening twilight, setting only an hour after the Sun by the 30th. Moon nearby on the 12th.

JUPITER, also in Gemini, moves into NW evening twilight setting an hour after sunset by the 30th. Moon nearby on the 12th.

SATURN is in conjunction with the Sun on the 9th and then becomes a morning object, though not visible in June.

URANUS is in Aquarius rising by 22h 30m by the 30th. Stationary on the 3rd. Moon to the south on the 1st and 29th.

NEPTUNE is in Capricornus rising by 22h by the 30th. Moon to the north on the 27th.

The Moon

Last quarter 3d 00h
New Moon 11d 00h
First quarter 18d 00h
Full Moon 24d 22h

Eclipses on the 11th and 24th: *see* page 3
The Earth: Summer Solstice 21d 13h

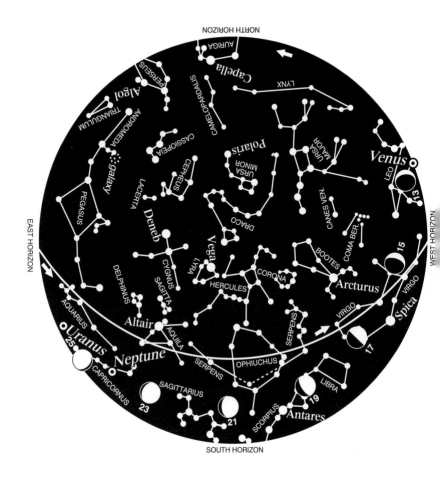

JULY 1, 23h (11pm)

The aspect of the sky (apart from the Moon and Planets)
will be approximately the same in other months at the
following times:

**April 1, 05h: May 1, 03h: June 1, 01h:
August 1, 21h: September 1, 19h.**

The time in these notes is that of the Greenwich meridian.

JULY

The Planets

MERCURY might be glimpsed very low in NE morning twilight during the first week. In the evening sky after superior conjunction on the 21st it sets too soon after the Sun to be visible.

VENUS is a brilliant –4.1 magnitude evening object in NW twilight, setting 2h after the Sun by the 31st. Moon nearby on the 13th. Regulus close below Venus on the 10th.

MARS is 1.8 magnitude and in bright twilight, setting only minutes after the Sun by the 31st. Jupiter close by on the 2nd–3rd.

JUPITER is in conjunction with the Sun on the 20th and then moves into the morning sky though not observable in July.

SATURN is 0.0 magnitude and in Taurus rising by 01h by end July. Moon nearby on the 8th.

URANUS is in Aquarius rising soon after sunset by the 31st. Moon to the south on the 26th.

NEPTUNE is in Capricornus rising shortly after sunset by the 31st. Moon to the south on the 24th.

The Moon

Last quarter 2d 17h
New Moon 10d 10h
First quarter 17d 05h
Full Moon 24d 09h

The Earth: at aphelion 6d 04h (152 million km)

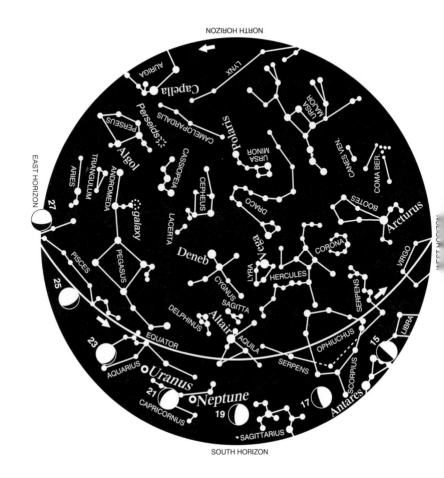

AUGUST 1, 23h (11pm)

The aspect of the sky (apart from the Moon and Planets) will be approximately the same in other months at the following times:

June 1, 03h: July 1, 01h: September 1, 21h: October 1, 19h: November 1, 17h.

The time in these notes is that of the Greenwich meridian.

AUGUST

The Planets

MERCURY is an evening object but will not be observable in August.

VENUS is at greatest eastern elongation (46 degrees) on the 22nd but being well south of the Sun will set only 1h after sunset by the 31st. The –4.4 magnitude planet should be visible very low in the west. Moon nearby on the 11th.

MARS is in conjunction with the Sun on the 10th and will not be observable this month.

JUPITER is –1.8 magnitude and in Cancer rising about 02h by the 31st.

SATURN is 0.1 magnitude and passes from Taurus into northern Orion rising about 23h by end August. Moon nearby on the 4th.

URANUS is in Aquarius moving back into Capricornus and at opposition on the 20th. Moon to the south 22nd.

NEPTUNE is in Capricornus and at opposition on the 2nd. Moon to the south 20th.

The Moon

Last quarter 1d 10h
New Moon 8d 19h
First quarter 15d 10h
Full Moon 22d 22h
Last quarter 31d 03h

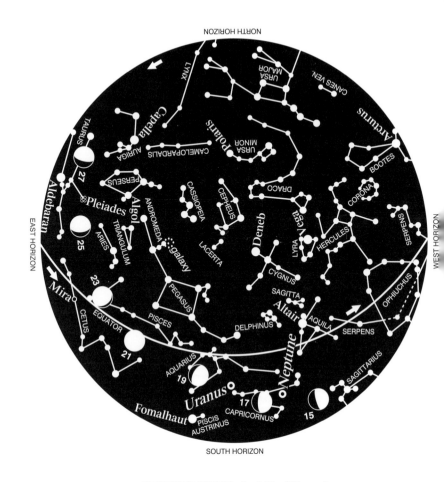

SEPTEMBER 1, 23h (11pm)

The aspect of the sky (apart from the Moon and Planets) will be approximately the same in other months at the following times:

**July 1, 03h: August 1, 01h: October 1, 21h:
November 1, 19h: December 1, 17h.**

The time in these notes is that of the Greenwich meridian.

SEPTEMBER

The Planets

MERCURY is at greatest eastern elongation on the 1st (27 degrees) and at inferior conjunction on the 27th but will not be visible this month.

VENUS sets less than an hour after the Sun this month but at greatest brilliancy of –4.5 magnitude on the 26th may be visible low in the west after sunset.

MARS is now a morning object in Leo but will be in twilight during September.

JUPITER is in Cancer and –2.0 magnitude rising by 01h by the 30th. Moon nearby on the 4th–5th.

SATURN is in Orion and rises by 21h 30m by the 30th. Moon nearby on the 1st and 28th.

URANUS is in Capricornus and sets about 02h by the 30th. Moon to the south on the 18th.

NEPTUNE is in Capricornus and sets about 01h by the 30th. Moon to the south on the 17th.

The Moon

New Moon 7d 03h
First quarter 13d 18h
Full Moon 21d 14h
Last quarter 29d 17h

The Earth: Autumn Equinox 23d 05h

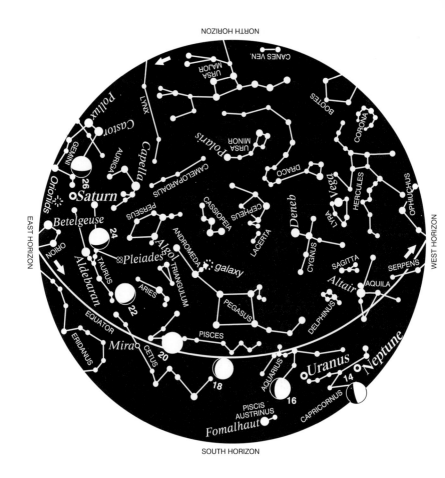

OCTOBER 1, 23h (11pm)

The aspect of the sky (apart from the Moon and Planets) will be approximately the same in other months at the following times:

August 1, 03h: September 1, 01h: November 1, 21h: December 1, 19h: January 1, 17h.

The time in these notes is that of the Greenwich meridian.

OCTOBER

The Planets

MERCURY is at greatest western elongation on the 13th (18 degrees). It rises 2h before the Sun in mid-month and brightens to –1.0 magnitude by the 31st. It should be visible in eastern morning twilight during the last two weeks of October.

VENUS will be lost in evening twilight to be at inferior conjunction on the 31st.

MARS is 1.8 magnitude, passing from Leo into Virgo and can now be seen in a dark sky rising about 04h by the 31st. Moon nearby on the 5th.

JUPITER is –2.0 magnitude, in Cancer and rises about 23h by the 31st. Moon nearby on the 1st–2nd and 29th.

SATURN is in Orion and rises about 19h by the 31st. It is stationary on the 11th. Moon close by on the 25th.

URANUS is in Capricornus and sets about 0h by the 31st. Moon to the south on the 15th–16th.

NEPTUNE is in Capricornus and sets about 22h 30m by the 31st. Stationary on the 20th. Moon to the south on the 14th.

The Moon

New Moon 6d 11h
First quarter 13d 06h
Full Moon 21d 07h
Last quarter 29d 05h

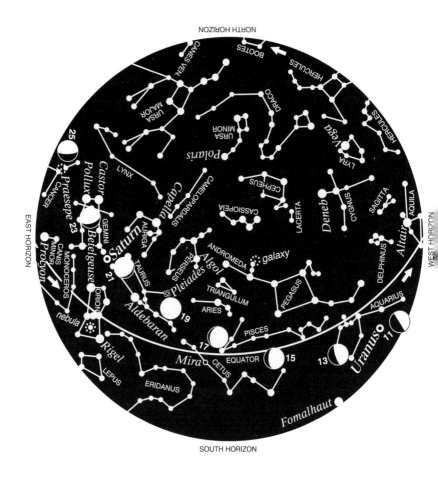

NOVEMBER 1, 23h (11pm)

The aspect of the sky (apart from the Moon and Planets)
will be approximately the same in other months at the
following times:

**September 1, 03h: October 1, 01h: December 1, 21h:
January 1, 19h: February 1, 17h.**

The time in these notes is that of the Greenwich meridian.

NOVEMBER

The Planets

MERCURY reaches superior conjunction on the 14th and then becomes an evening star but not observable this month.

VENUS draws rapidly away from the Sun to rise 3h before sunrise by the 30th. The –4.4 magnitude inner planet should be visible in the SE dawn sky from mid-month.

MARS is in Virgo rising by 04h by the 30th. Mars N of Spica on the 20th. Moon nearby on the 2nd.

JUPITER passes from Cancer into Leo, rising by 21h 30m by end month. Moon nearby on the 25th–26th.

SATURN passes from Orion back into Taurus, rising an hour after sunset by the 30th to be above the horizon all night. Moon nearby on the 22nd.

URANUS is in Capricornus and sets by 22h by the 30th. Stationary on the 4th. Moon to the south on the 11th.

NEPTUNE is in Capricornus and sets about 20h 30m by the 30th. Moon nearby on the 10th.

The Moon

New Moon 4d 21h
First quarter 11d 21h
Full Moon 20d 02h
Last quarter 27d 16h

Eclipse on the 20th: *see* page 3

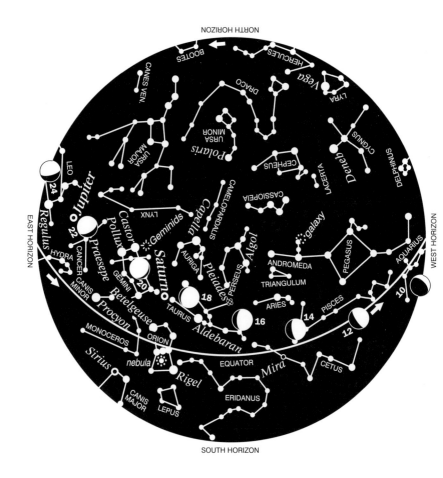

DECEMBER 1, 23h (11pm)

The aspect of the sky (apart from the Moon and Planets)
will be approximately the same in other months at the
following times:

**September 1, 05h: October 1, 03h: November 1, 01h:
January 1, 21h: February 1, 19h.**

The time in these notes is that of the Greenwich meridian.

DECEMBER

The Planets

MERCURY is an evening star and should become visible low in the SW during the last week of December. It reaches greatest western elongation (20 degrees) on the 26th when it will be –0.5 magnitude.

VENUS is at greatest brilliancy on the 7th, a –4.6 magnitude morning star, rising by 04h. Moon nearby on the 1st and 30th.

MARS is 1.6 magnitude passing from Virgo into Libra and rising about 04h during December. Moon nearby on the 1st and 30th.

JUPITER is –2.5 magnitude and moves back from Leo into Cancer, rising by 19h 30m by the 31st. Stationary on the 4th. Moon nearby on the 23rd.

SATURN –0.4 magnitude and at opposition in Taurus on the 17th when it will be above the horizon all night. It moves from Orion back into Taurus in December. Moon nearby on the 19th.

URANUS is in Capricornus and sets soon after 20h by the 31st. Moon nearby on the 9th.

NEPTUNE is in Capricornus and sets about 19h by the 31st. Moon nearby on the 8th.

The Moon

New Moon 4d 08h
First quarter 11d 16h
Full Moon 19d 19h
Last quarter 27d 01h

Eclipse on the 4th: *see* page 3.
The Earth: Winter Solstice 23d 01h

THE STARS

The stars are subdivided into magnitudes according to apparent brightness; the lower the number the brighter the star and the larger the dots on our monthly maps. Any star is about 2½ times as bright as one of the next magnitude. The faintest star ordinarily visible to the naked eye is of the 6th magnitude, or just one-hundredth of the brightness of one of the 1st, but that is possible only under a very clear sky. On a moonless night the total number of stars so visible is about 1,000. The faintest object detected with ground-based telescopes is of the 28th magnitude though the Hubble Space Telescope has now reached 30th magnitude.

Zero magnitude (0.0) represents a brightness 2½ times that of a standard first-magnitude star. Brightnesses in excess of this are indicated by a minus sign, the magnitude of Sirius, for example, being -1.47. Venus at its brightest is -4.6 or 145 times as bright as a first-magnitude star. The magnitude of the Full Moon is -12.5, equal to 250,000 first-magnitude stars. The stellar magnitude of the Sun is -26.6 or some 444,000 Full Moons.

The colours of the stars are indications of their surface temperatures. The temperature of a reddish star like Antares is about 3,000°C, and that of a bluish-white star, such as Vega, is about 11,000°C. The temperatures of orange, yellow and white stars are intermediate between these extremes.

An examination of the sky on a clear dark night shows that the distribution of stars is far from uniform. While there are distinct clusters of stars, such as the Pleiades and Praesepe, many other groupings consist of stars that just happen to lie in the same direction but at very different distances. The most noticeable concentration of stars is towards what we call the Milky Way, the faint band of light that passes through the following constellations: Puppis, Monoceros, Gemini, Auriga, Perseus, Cassiopeia, Cepheus, Cygnus, Aquila and Sagittarius. Not all of these constellations are above the horizon at any one time. The band of the Milky Way actually extends right round the sky passing through some southern constellations that never rise above the horizon in the British Isles.

Even binoculars show that the Milky Way is made up of thousands of stars, too faint to be seen with the naked eye. Our Sun is situated well away from the centre of a huge, flattened, disc-like system of stars 100,000 light years across called the Galaxy. It contains more than 100,000 million stars. When we look along the plane of the disc we see the star-clouds of the Milky Way; but when we look out above or below the plane we see far fewer stars.

From a distance, the Galaxy would look like that in Andromeda, visible to the naked eye only as a hazy oval patch of light. This is one of the nearer galaxies, only two-million light years away. Others have been found in their millions; some may be farther than 10,000 million light years distant, each containing thousands of millions of stars. The central bulge of our Galaxy lies towards the great star clouds in Sagittarius, not easily seen from our latitudes.

SUNSET, SUNRISE AND NAUTICAL TWILIGHT

Date		London Area				Edinburgh Area			
		Sunset	End NT	Begin NT	Sunrise	Sunset	End NT	Begin NT	Sunrise
Jan	1	16 00	17 20	06 45	08 08	15 45	17 22	07 08	08 44
	15	16 20	17 40	06 40	08 00	16 10	17 43	07 00	08 45
Feb	1	16 45	18 05	06 25	07 40	16 41	18 10	06 40	08 10
	15	17 15	18 30	06 00	07 15	17 16	18 38	06 15	07 38
Mar	1	17 35	18 50	05 35	06 50	17 47	19 05	05 40	07 03
	15	18 05	19 15	05 00	06 15	18 15	19 35	05 05	06 28
Apr	1	18 35	19 50	04 20	05 35	18 49	20 12	04 20	05 45
	15	19 00	20 10	03 40	05 00	19 20	20 52	03 45	05 05
May	1	19 25	20 50	03 05	04 30	19 50	21 35	02 45	04 25
	15	19 45	21 35	02 30	04 05	20 15	22 25	01 56	03 47
Jun	1	20 10	22 00	01 55	03 50	20 43	23 53	00 35	03 34
	15	20 20	22 29	01 33	03 40	21 00	-----	-----	03 25
Jul	1	20 25	22 25	01 40	03 45	21 01	-----	-----	03 31
	15	20 10	22 00	02 05	04 00	20 45	23 40	00 45	03 45
Aug	1	19 50	21 30	02 40	04 20	20 19	22 22	02 10	04 15
	15	19 25	20 50	03 15	04 45	19 50	21 30	03 00	04 42
Sep	1	18 50	20 10	03 50	05 10	19 09	20 42	03 45	05 15
	15	18 15	19 30	04 20	05 35	18 30	19 52	04 20	05 43
Oct	1	17 40	18 50	04 45	06 00	17 47	19 07	04 55	06 15
	15	17 05	18 20	05 10	06 25	17 05	18 32	05 25	06 40
Nov	1	16 35	17 50	05 40	06 50	16 31	17 57	05 52	07 19
	15	16 10	17 25	06 02	07 20	16 04	17 30	06 22	07 48
Dec	1	15 50	17 15	06 25	07 45	15 42	17 16	06 45	08 20
	15	15 50	17 13	06 37	08 03	15 36	17 10	07 00	08 38
	31	16 00	17 20	06 45	08 08	15 45	17 22	07 08	08 44

Times in UT. Also see notes overpage

Notes:

1. Times are given in Universal Time (= GMT): when British Summer Time (BST) is in force (usually from late March to late October) add 1 hour.

2. Nautical Twilight is defined as the moment when the Sun's true centre reaches a depression of 12 degrees below the horizon. Then it is dark enough to see the brighter stars and planets, and in suburban areas it often gets no darker. When no time is shown nautical twilight lasts all night.

3. Times given are approximate. They depend on the observer's latitude and longitude. Sunset, sunrise and twilight times will be 4 minutes earlier for every degree of longitude east of the Greenwich meridian and 4 minutes later for every degree west. London is on the Greenwich meridian; Edinburgh is about 3 degrees west or 12 minutes later.

4. The observer's latitude also affects these times: for example sunset occurs earlier in Edinburgh than London in winter but later in summer. It is not possible to cover more than two regions here, but a reasonable estimate can be made for other parts of the British Isles. The times may be used for any year. More explanation on this and other subjects can be found in *The Times Night Sky Companion*.

PRINCIPAL METEOR SHOWERS IN 2002

Name	Period of maximum visibility	Average hourly rate	Notes on visibility & moonlight
Quadrantids	2–4 Jan	10–100	Unfavourable. Last Qtr on 6th. Radiant low in N in evening
Lyrids	21–22 Apr	10	Good later. First Qtr on 20th
Perseids	11–14 Aug	60	Favourable. First Qtr on 15th
Orionids	20–22 Oct	10–20	Unfavourable. Full on 21st
Taurids	Late Oct–late Nov	5–10	Slow meteors from below Pleiades
Leonids	16–18 Nov	??	Strong shower possible this year. Unfavourable, full on 20th
Geminids	12–14 Dec	60	Good after 01h., First Qtr on 11th

Notes: The Leonids are normally a weak shower but every 33 years activity increases when the parent comet Tempel-Tuttle is near the Sun. Then high but usually short-lived activity may be seen as in the display of fireballs in 1998 and high rates of fainter meteors in 1999. The radiant, from where the meteors appear to come, is from within the 'sickle' (or reversed question mark) of the constellation Leo (major): *see* the December chart (p.26).

Moonlight between first and last quarter seriously interferes with the number of faint meteors seen. Fewer meteors are usually seen when the radiant area is low near the horizon. Radiant areas are shown on the monthly charts, except for the Taurids which covers a wide area and the Leonids which has not risen at the time of the November chart.

Further Events in 2002

The monthly notes give the magnitudes of the planets as a guide as to how bright they will appear (also *see* p.28). Venus and Jupiter are always so bright that they are easily identified but Mars and Saturn may be confused with nearby bright stars. Mercury is visible only during twilight. The following easily found bright stars, spread throughout the year, may help to identify the planets by comparing their brightness (magnitude): Polaris (2.0), Aldebaran (0.9), Sirius (–1.5), Castor (1.6), Procyon (0.4), Regulus (1.3), Arcturus (0.0), Spica (1.0) and Altair (0.8). (See page 28 for an explanation of magnitudes.) For morning events the time of the beginning of nautical twilight (BNT) is usually used as a guide to the time when it becomes too light for further observation. The end of nautical twilight (ENT) is used in the evening for when it becomes dark enough. The times below are for London and will vary elsewhere (*see* p.30).

Various groupings and planetary conjunctions of Mercury, Venus, Mars, Saturn, Jupiter and the Moon take place in the evening sky from mid-April until the end of June. Jupiter passed Saturn (now in Taurus) in 1999 and is now well to the east in Gemini so any conjunctions cannot involve these two planets.

The outer planets move more slowly than the Earth and have larger orbits so they all take longer than a year to orbit the Sun. Mars takes 1.9 years, Jupiter 12, Saturn 29.5, Uranus 84 and Neptune 165 years. So Jupiter moves eastwards against the stars by 30 degrees each year while Saturn moves only 12 degrees. Saturn will stay over twice as long as Jupiter in any constellation. While Mars is making one orbit of the Sun, the Earth will have made nearly two and it will be 2 years and 2 months before the planets again line up with the Sun and Mars again comes to opposition. After opposition Mars remains in the evening sky for many months, setting about the same interval after sunset, until eventually it is overtaken by the Sun and is lost in evening twilight. So we get to see Mars well every two years for a few weeks and when it is near opposition it can rival Jupiter in brightness. Then there are long periods when the planet is in evening and the morning twilight, rather faint and inconspicuous, before the next close opposition and approach to the Earth.

The inner planets Mercury and Venus also move eastwards with the Sun but swing to the east and west of the Sun as they become evening or morning objects. Mercury takes 88 days to orbit the Sun and Venus 225 days but their movements are more complicated as being in orbits smaller than that of the Earth, they can never appear to us to be very far from the Sun; for Mercury the maximum distance is 28 degrees and for Venus 47 degrees. In a year Mercury usually has three morning apparitions with another three in the evening. Generally only one or two are favourable to observers at our latitude. Venus takes 19 months from one superior conjunction to the next. More information on the planets and their movements (and other naked eye phenomena) can be found in *The Times Night Sky Companion.*

This year fast-moving Venus and Mars move eastwards to overtake first Saturn and then some weeks later Jupiter. The timings of the more spectacular events follow.

23 February: At about 02h 45m the waxing gibbous Moon will occult Jupiter with the planet reappearing at the bright western limb about 40 minutes later, by which time they will be slipping down towards setting at the NW horizon.

16 April: at ENT or 20h 20m in London, the planets will line up along the ecliptic with Venus nearest the western horizon, then Mars near the Pleiades, the Moon and Saturn very close, and Jupiter above-left in Gemini. At about 21h the crescent Moon will pass in front of Saturn (ie. an occultation of Saturn by the Moon). Saturn will disappear at the Moon's dark eastern limb passing behind the southern part of the Moon's disc to reappear about 20 minutes later at the bright western limb. The event can be seen best in a telescope or in binoculars.

Note: for the May–June events the planets are named in order of their distance away from the western horizon, that is upwards and towards the left.

1–14 May: by the beginning of May ENT will be about 21h when the planets will be nearer the NW horizon than in mid-April. Mercury, Venus, Mars and Saturn will be spread over about 10 degrees. By the **6th May** Venus, Saturn (below) and Mars will form a triangle with Mercury away to the NW (right). On the **10th May** Venus passes just to the north of Mars with Saturn below and Mercury at about the some altitude to the NW. By the **14th May** Mercury will still be to the right of Saturn but by ENT they will both be almost on the horizon even from southern England and probably no longer visible. Mars, the crescent Moon and Venus will form a compact triangle less than 10 degrees above the NW horizon, with Jupiter above left in Gemini.

3 June: by early June only Venus and Jupiter will be left in the evening sky and they will be very low in the NW at 22h 10m (ENT). In Gemini, Venus will pass just to the north of Jupiter. On the **13 June** at 21h 30m, with ENT not until 22h 25m, Mars and Jupiter, although just above the NW horizon, will be lost in twilight but Venus just to the south of the crescent Moon should be visible, in line with Castor and Pollux.

3 July: Mars passes to the north of Jupiter though the planets will be in too bright a western sky for this to be observable.